DNA

Rebecca Woodbury, Ph.D., M.Ed.

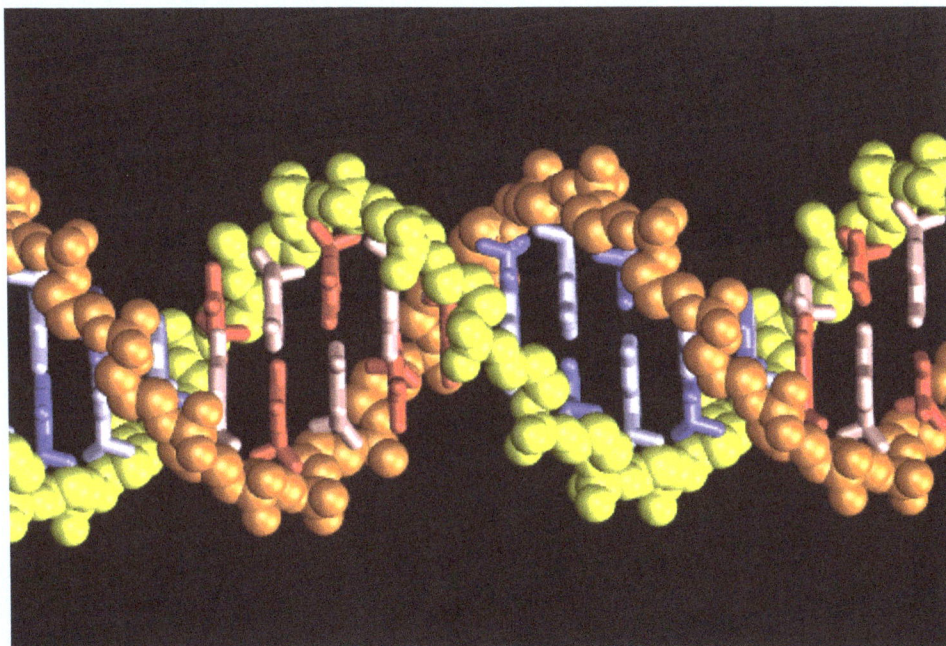

Gravitas Publications Inc.

DNA

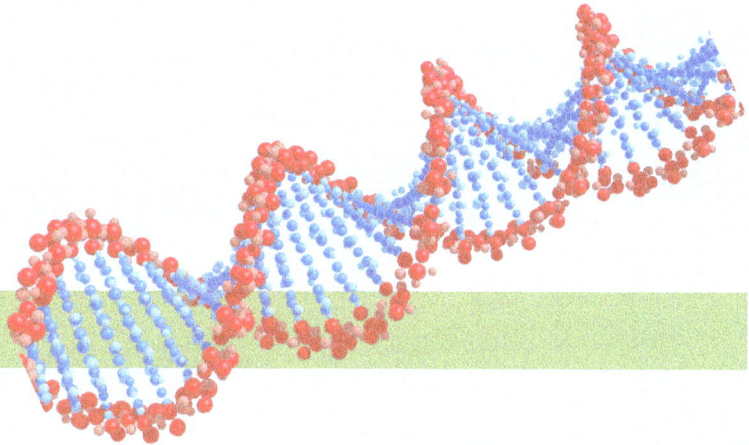

Illustrations: Janet Moneymaker

DNA
ISBN 978-1-950415-17-5

Published by Gravitas Publications Inc.
Imprint: Real Science-4-Kids
www.gravitaspublications.com
www.realscience4kids.com

RS4K

Photo credits: Cover, Title Pg, P. 15: Illustration by D. J. Keller; Above & P. 7, Image by Alexander Antropov from Pixabay

Have you ever watched a crime story?

Did you hear someone say **DNA?**

Do you know how to say DNA?

DEE - EN - A

But what exactly is DNA?

What is it?

I don't know.

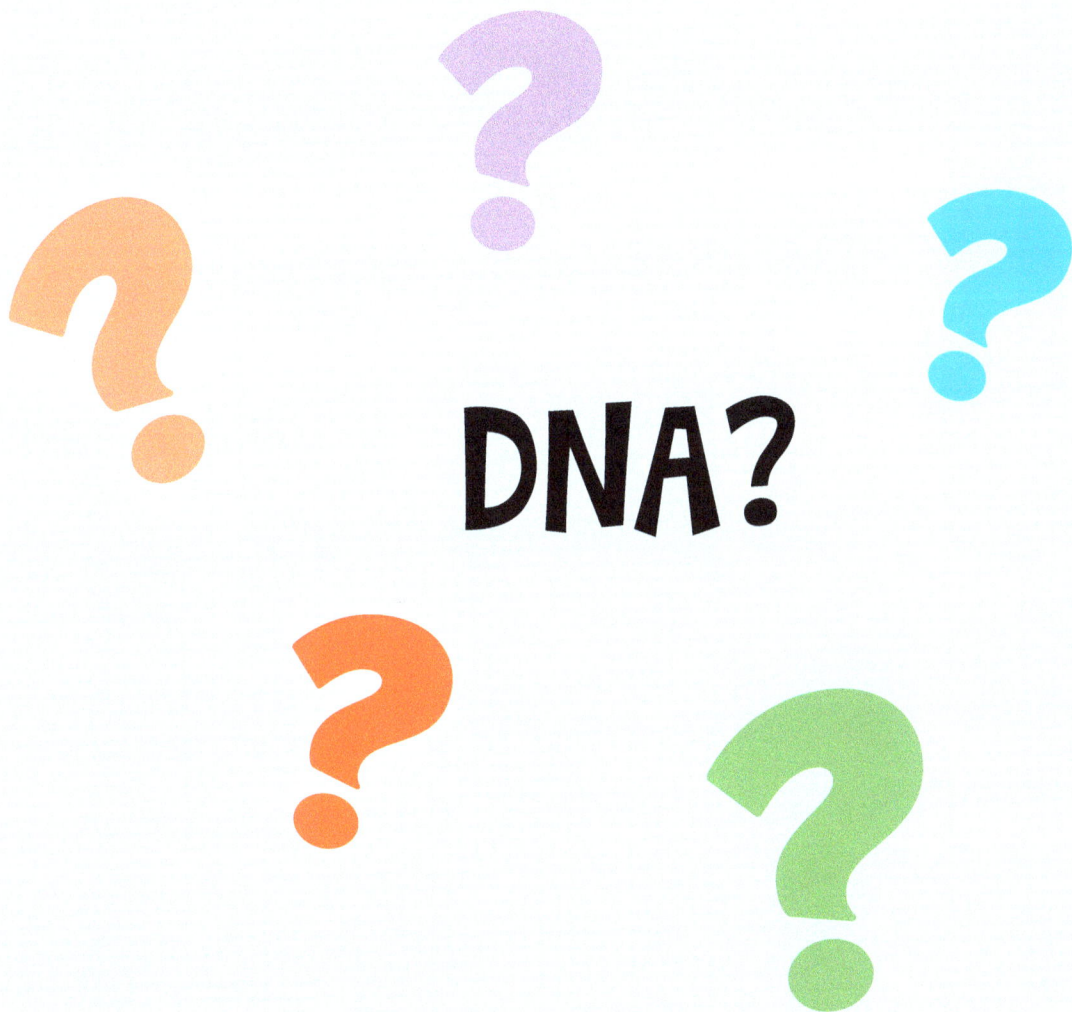

DNA?

- **DNA** is found inside the cells of living things.

- **DNA** stands for **deoxyribonucleic acid.**

- **DNA** is a **polymer** made of two long stands. Each strand wraps around the other.

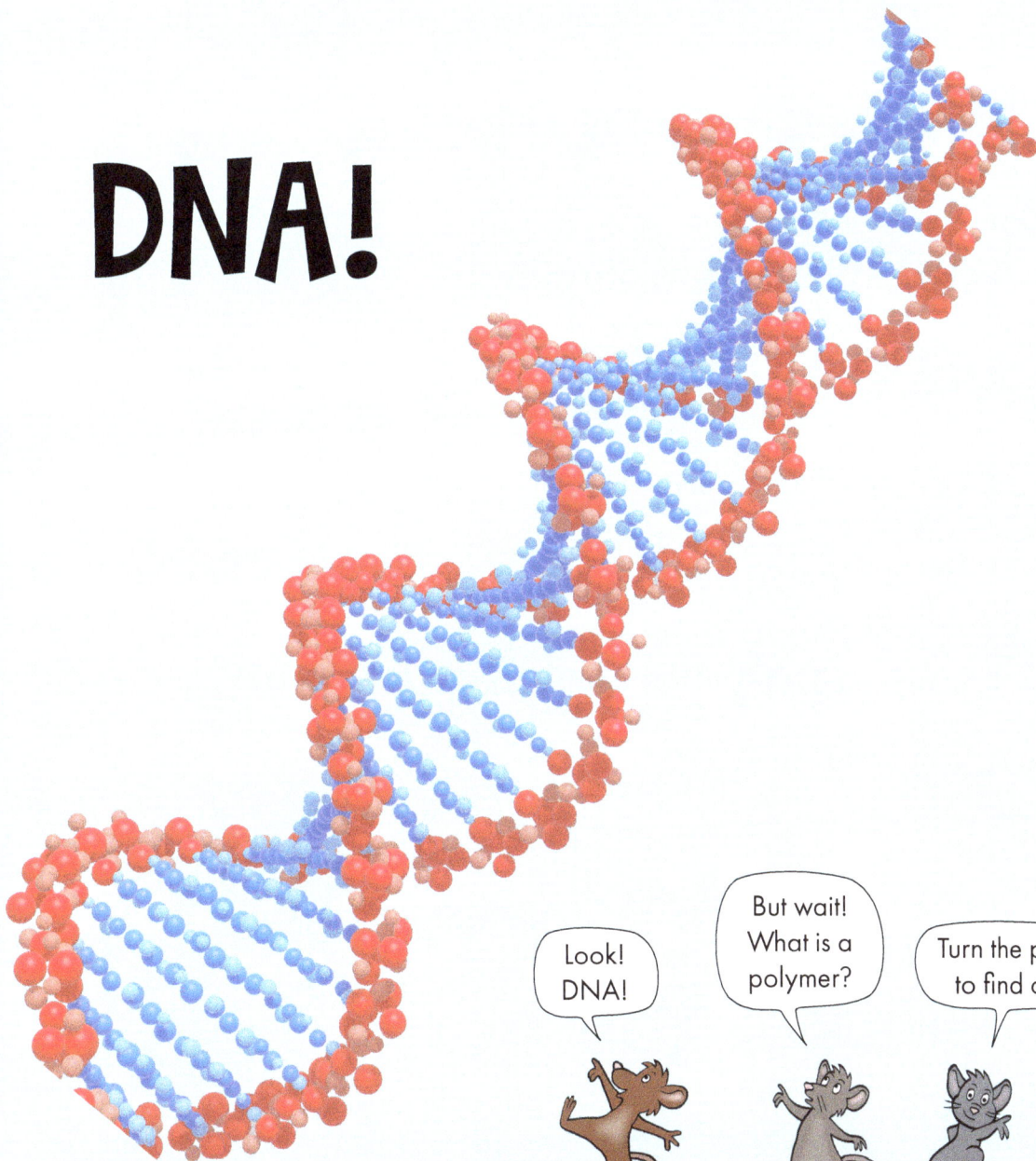

DNA!

Look! DNA!

But wait! What is a polymer?

Turn the page to find out.

Review: POLYMER

A **polymer** is a **molecule.**

It is a long chain of atoms linked together.

"Poly" means "many."

"Mer" means "unit."

The word **polymer** means **many units.**

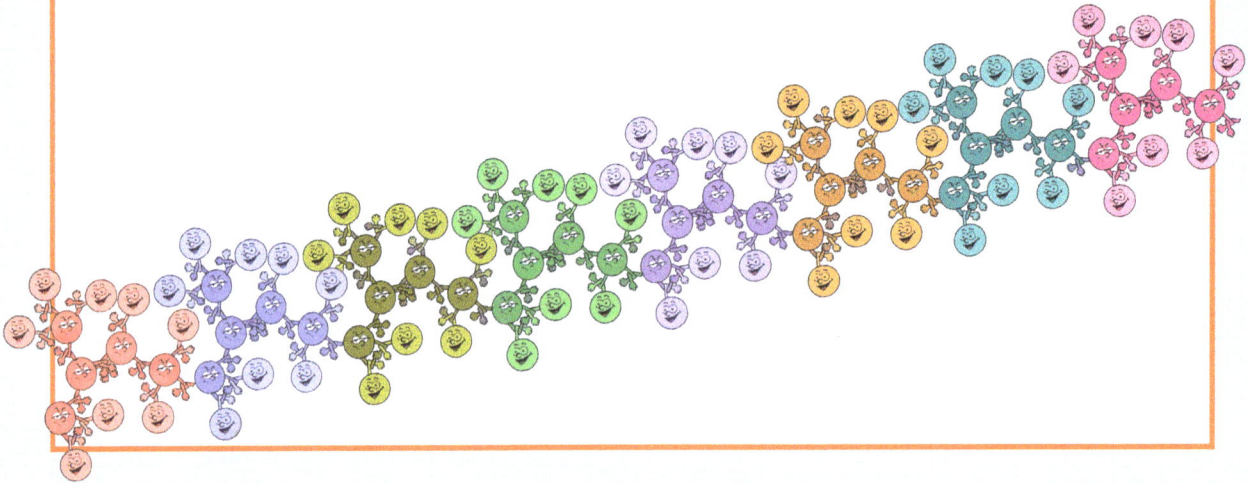

Review: ATOMS

Atoms are tiny building blocks that can link together.

Atoms make everything we see, touch, taste, and smell.

Review: MOLECULES

Molecules are made when **atoms link** together.

DNA looks like a twisted ladder.

This shape is called a **double helix.**

It is **double** because it has two strands.

It is a **helix** because it makes a coil.

Does my tail make a helix?

No.

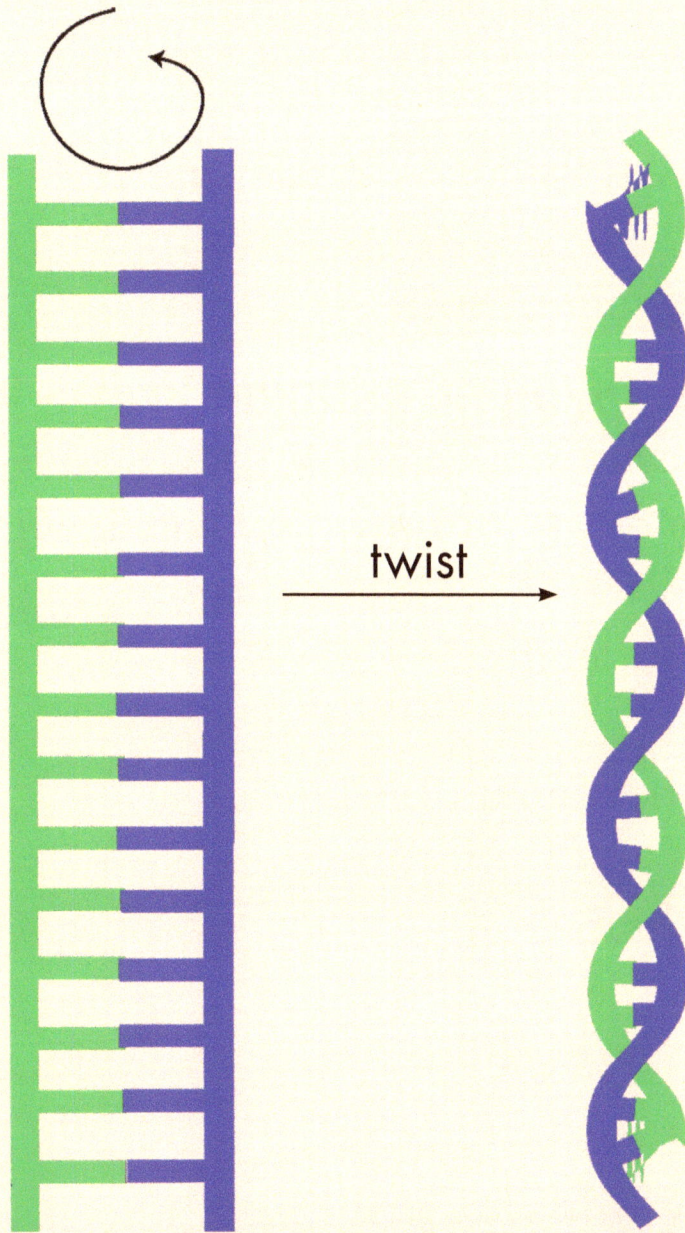

twist

The rungs of the DNA ladder are called **bases.**

There are four bases:
adenine, thymine, guanine, and **cytosine.**

Their names are abbreviated with letters:

A (adenine)

T (thymine)

G (guanine)

C (cytosine)

DNA Bases

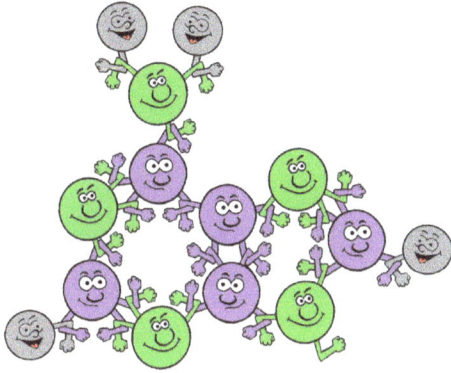

Adenine
A

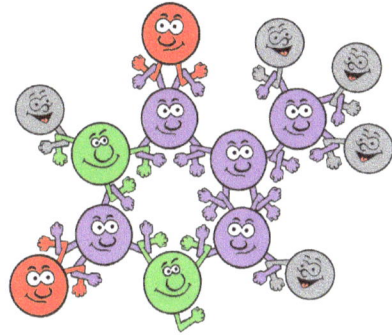

Thymine
T

Guanine
G

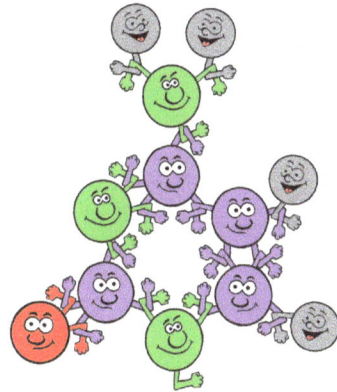

Cytosine
C

The bases line up along the inside of the double helix.

A pairs with **T**.

G pairs with **C**.

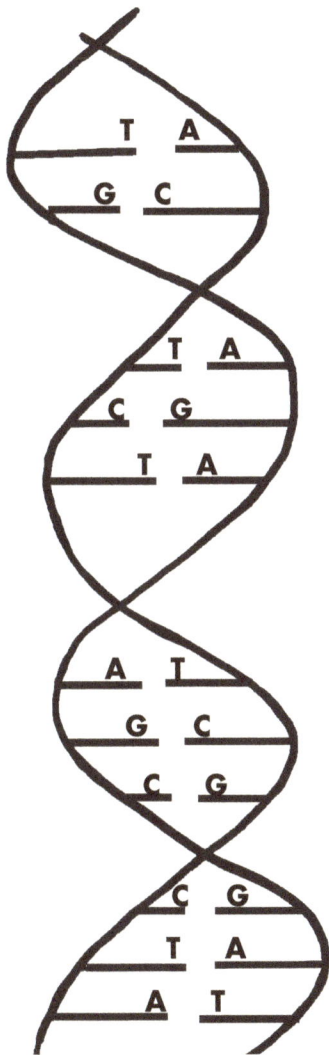

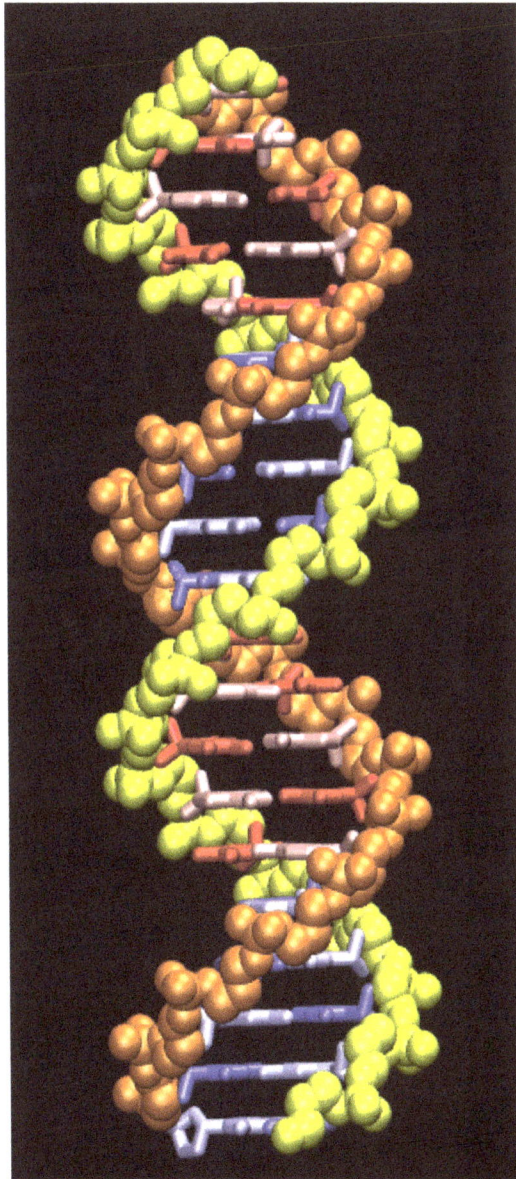

The order of the bases along the DNA strands makes a **code.**

This code is read by **proteins.**

Review: PROTEINS

Proteins are molecules made of long strands of linked atoms.

Proteins fold into many different shapes and sizes.

Proteins do special jobs inside cells.

The DNA code is like a blueprint or map for your body.

The DNA code is called the **genetic code.**

All living things have a genetic code.

Did you know you have a genetic code?

I do now!

A **genetic code** determines if you will have brown hair or blond hair. It determines if you will be short or tall and if you will have dark skin or light skin.

You get your genetic code from your parents. Each person has a different and unique genetic code.

But you are more than just your DNA and genetic code. What you eat, where you live, and what you do help make you unique in every way.

How to say science words

adenine   (AA-duh-neen)

atom   (AA-tum)

cytosine   (SIY-tuh-seen)

deoxyribonucleic acid
   (dee-AHK-see-riy-boh-new-klay-ik   AA-sed)

DNA   (DEE - EN - A)

double helix   (DUH-buhl   HEE-liks)

genetic code   (juh-NE-tik   COHD)

guanine   (GWAH-neen)

molecule   (MAH-lih-kyool)

polymer   (PAH-luh-muhr)

protein   (PROH-teen)

thymine   (THIY-meen)

www.ingramcontent.com/pod-product-compliance
Lightning Source LLC
Chambersburg PA
CBHW040149200326
41520CB00028B/7541